Quantum Mechanics A - Z without the BS

Quantum Mechanics A - Z without the BS

Al Schneider

Cover by Al Schneider
Art by Al Schneider
Custom typesetting software by Al Schneider
HTML to PDF by HTMLDOC: Easy Software Products

Contents

Foreword

The goal of this book is to explain quantum mechanics in a direct and efficient manner. Hopefully, you will get a gut level understanding of what quantum mechanics is and why it is important to our lives. While little math is used, some understanding of contemporary science is assumed. For example, this work assumes you know what electrons, protons, photons, and neutrons are. This little book is a subset of a book titled, New Age Quantum Physics, which details the history of quantum physics and explains the phenomena presented here.

Introduction

Quantum mechanics is about observations made during the past 2000 or more years. Often people focus on the odd things about the subject and the odd ideas that misinformed people say about it. This book reviews the critical observations of the subject and focuses on real world implications. The middle of this book pulls results of the observations together to get a unified idea of what quantum mechanics is. Finally, these concepts are used to explain how transistors function.

Aristotle

The first observation we notice in recorded history is by Aristotle over 2000 years ago. He proposed that light was a wave. No doubt, he observed water waves and sound waves. He must have head a sound from afar and compared that to a flash of light some distance away as a ray of sunlight bounced off something in the distance.

Huygens and Young

The next important observations consisted of comparing water waves to light waves. Both display diffraction and interference. This added strength to Aristotle's claim that light was a wave.

Maxwell's Equations

In somewhat recent history, several men studied electricity and the effects it produced on the area around electrons. A man named Maxwell gathered their work expressed as a set of mathematical equations. These are referred to as Maxwell's equations. One of the results of this study was an equation describing the space through which light passed. The equation was similar to the equation of water waves in Huygens's and Young's experiments.

The Nucleus is Discovered

In 1911, Rutherford ran an experiment in which high velocity particles radiating from a piece of radium were aimed at a thin piece of gold foil. The conclusion of the experiment was that the atom had a very hard center with electrons whirling about that center.

Birth of the Quantum

Planck investigated the known observation that hot metal radiated different colors of light relative to the temperature of the metal. When measuring the energy of the atoms in hot metal, he noticed the light given off appeared with fixed frequencies and depended directly on the temperature of the metal. He discovered the differences in associated energies varied by some small constant amount. This came to be called a quantum. The constant amount became known as Planck's constant. This marked the beginning of the quantum age.

Is Light a Particle or a Wave?

Then Einstein observed that light must consist of particles when he explained the photoelectric effect.

Is Matter Made of Waves?

Then a Frenchman named de Broglie suggested that particles are waves. An experiment by Davisson-Germer in 1927 observed that this was true.

The Birth of Uncertainty

Finally, Heisenberg, through a complicated series of mathematical calculations, observed the Heisenberg uncertainty principle.

Each of these is a fascinating subject to study. This book attempts to show how each of these phenomena were

observed and points out that mathematics was developed
to measure the phenomena and predict how the phenomena
would behave in applications that would be useful to man.
That is, how does each of these observations work
together to enable us to make something like transistors.
To meet this need, the second half of the book uses
information from the first half to describe how a transistor
works.

Chapter 1

Water and Light Waves

Over 2000 years ago, Aristotle wrote that light was a wave like the waves of the ocean. [1] He proposed that light traveled on some fine material that permeated the universe called ether. The concept of ether was accepted for over 2000 years until it was recently challenged and found to be wanting.

From Aristotle we leap forward in time to review three men and their ideas about light. They were Huygens, Newton, and Young. Huygens presented his theory in 1678. Newton presented his in 1704. Young followed in 1801. Huygens presented a wave theory that was for the most part ignored. It appeared 26 years before Newton's. [2] Then, during Newton's time, due to his stature in the physics community, Huygens' theory was totally discarded. Newton proposed what he called a corpuscular theory of light. It described light as particles. Newton's theory reigned for some 150 years. It fell from favor, as it did not explain diffraction and interference. [3] Young came along with his experiment about 100 years later. [4] It captured attention but fell short due to a few small problems and the stature of the Newton image. Some of the problems with Young's theory were resolved about 50 years later and Young's theory reigned until the emergence of the quantum age.

Even then, Young's theory was not rejected. Rather, it became part of the mix of quantum weirdness. Both Huygens and Young's experiments remain important factors in the quantum age. Although Newton's theories of light were closer to some concepts of the new age of quantum physics, his ideas are not close enough to catch on.

Huygens' One Slit Experiment

In 1678, Huygens observed diffraction in water waves and light. To him, both phenomena appeared the same. The following depicts the diffraction of water waves.

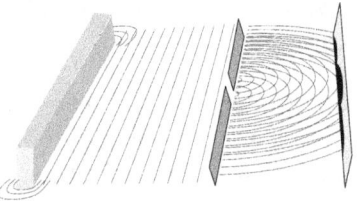

The wood block on the left of the picture moves. The motion creates water waves that move to the right. There is a barrier with a small opening in it, which only allows a small section of water waves to slip through. That generates circular waves that interact with each other. The dark areas on the barrier on the right have depicted where the water waves converge and splash water higher. [5]

The following depicts light waves doing the same thing. Here, however, we cannot see the light waves moving through the air. The light is forced to pass through a similar opening as in the previous picture. Here, the light waves form a pattern similar to the water waves on the barrier on the right. The white areas represent where the light strikes on this barrier. [5]

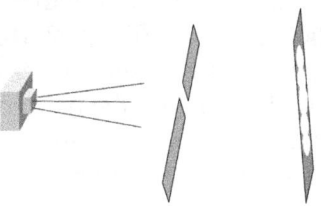

Both experiments produce the same result. The size of the opening in the barriers was about half the wavelength of water waves and presumed to be the same for light waves.

Light, Huygens suggested, consisted of the longitudinal vibrations of an all pervasive ether composed of small, hard, elastic particles, each of which transmitted the impulses it received to all contiguous particles. [6] According to this theory, the particles did not move but remained in a given position.

This presented a bit of a problem at the time as this back and forth motion would not support polarization. Water waves do not display an analogy to polarization. Nor do sound waves. Yet, light is polarized. However, the real barrier to the acceptance of Huygens' idea was the popularity of Isaac Newton and his concepts.

The Young Double Slit Experiment

Young's double slit experiment appears very similar to Huygens' experiment. It has another slit for the waves to go through. Here is how the experiment appears.

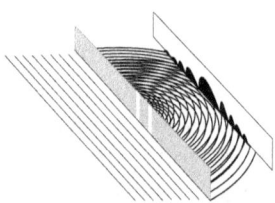

The double slits are close together. As the water waves move through the slits, they form two circular waves moving away from the slit barrier. They interfere to produce the interference pattern on the barrier on the right. Some water waves splash higher on the barrier to from the pattern shown in black. [7] The primary thing to notice is that the interference pattern is much higher than a pattern with a single slit.

Young observed the same behavior with light. [7]
Consider a similar experiment with light.

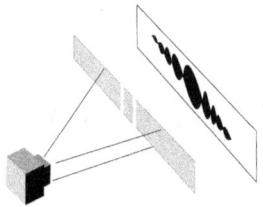

A pattern similar to water appears on the far barrier when light shines through two slits. As mentioned, there were a few problems with this theory of light. However, Fresnel and Foucalt resolved them about 50 years after Young introduced his theory. [8]

Why is this Important?

We, the public, are told that light is a wave. Looking at these experiments, we see that science has come to this conclusion because a light pattern on a wall produces a similar pattern as water waves when they splash against a wall. The key element to get from this discussion is the similarity between light and water. This remains a critical component of quantum physics today. Diffraction and interference became the definition of what a wave is.

Chapter 2

Maxwell's Equations

For the moment, we leave behind discussions of light and water waves. We need to take up a discussion of electricity moving through coils of wire. Scientists at the time noticed that when electricity flowed through a wire, the space around the wire affected things made of metal. They also noticed that when the wire was coiled about a wooden bobbin or perhaps a nail, the effect became stronger.

Maxwell and Friends

A man named Faraday in England studied the changes in that space and decided that the flowing electricity caused unseen lines of force in that space. In addition, Gauss, Lenz and Ampere as well as many others, studied these lines. Maxwell took some of Faraday's work and developed equations to express the concepts. Maxwell also gathered a number of equations from those other three men. These mathematical equations were gathered together in a package and popularized as Maxwell's Equations. They were published in a book called *On Physical Lines of Force*. [1]

This package is very complicated and beyond the math in this book. Nevertheless, we need to study one aspect of this package that has a great impact on the wave functions being studied here and ultimately on our study of quantum mechanics. This aspect can be introduced with an experiment using two wires. Consider the following.

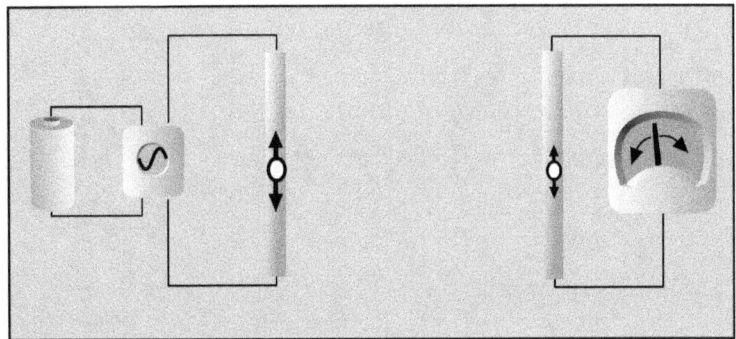

The left side of the picture above shows a battery attached to an oscillator that sends alternating current to a wire. The electrons in the wire first go up. Then the electrons in the wire go down. They continue this up and down motion. The wire on the left is enlarged to show an electron in the wire oscillating up and down. That is the dot in the middle of the left wire. On the right of the picture is another enlarged wire to show an electron in the wire. That wire is attached to a meter that can display the current in the wire. The needle is moving back and forth. There is no connection between the left wire and the right wire. However, the electron in the left wire moving up and down; causes the electron to move up and down in the right wire. This causes the needle in the meter to go back and forth. This demonstrates that an electromagnetic emission is traveling from the left to the right. There is nothing to see. This represents how radio and TV transmissions work.

When Maxwell's equations are applied to this problem, a

graphic image is produced showing how the space
between the two wires is affected.

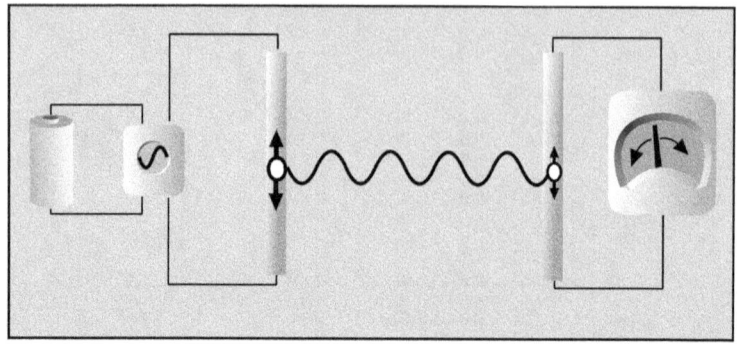

The squiggly line shows there is something in the space
between the wires that causes electrons to move up and
down. Maxwell's equations yield an equation for this
effect. The equation of this curve is the same as the water
wave equation and the equation used for light in the
diffraction and interference experiments. [2]

$$y = A \sin\left(2\pi\frac{x}{\lambda} - 2\pi\frac{t}{T}\right)$$

It means that when an electron is oscillating regularly, as
in the experiment, it somehow changes the environment
around it. The squiggly line in the picture represents the
change in the environment. It is said that the line
represents an E field or electric field. The various changes
in the squiggle line represent the strength of the E field in
that space. Where the curve is high, the E field is strong.
Where the field is low, the E field is weak. When a
charged particle such as an electron is immersed in such an
E field, there is a force exerted on the charged particle to
move up or down. That force causes the charged particle to
move, as depicted in these pictures. However, the E field
is oscillating. Therefore, the force on the particle is
oscillating and a charged particle immersed in it will move
up and down continuously.

Water vs. Electromagnetic Waves

The similarities between water waves and electromagnetic waves are staggering. The following picture puts them close to each other. The upper image represents a water wave with wooden blocks at each end of the wave. The lower image represents an electromagnetic wave with wires at each end of the wave.

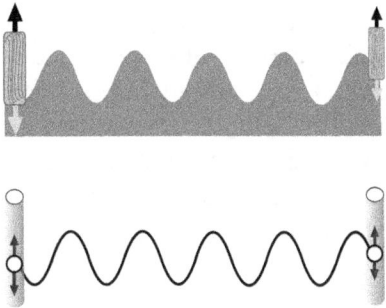

In the water wave, on the left, we have a block of wood moving up and down. In the electromagnetic wave, on the left, we have an electron moving up and down. Each has a wave that moves to the right. The water wave causes a block of wood to move up and down. The electromagnetic wave causes an electron to move up and down. Both have the same function or equation form.

$$y = A \sin\left(2\pi\frac{x}{\lambda} - 2\pi\frac{t}{T}\right)$$

Why is this Important?

The study of electrons moving through wires resulted in an equation describing something moving through the air from one wire to another. That equation is identical to the equation describing water waves and light in experiments done by Huygens and Young. To summarize, Maxwell's equations confirmed that light is a wave.

Chapter 3

The Biggest Scientific Myth

In 1911, science accepted a "plum pudding" model of the atom. In this model, bits of "plum" were electrons that were floating around in a "pudding" of positive charge. That year JJ Thompson decided to test that model. He is the man credited with discovering the electron. He recommended to one of his students, Ernest Rutherford, an experiment that would examine the pudding model of the atom.

The Rutherford Experiment

Rutherford designed an experiment that probed the depth of plum pudding idea. [1] The experiment used a high-energy source of alpha radiation. We know today that alpha radiation is two protons and two neutrons clumped together. This is the nucleus of a helium atom. Some radioactive substances constantly emit these particles as radioactive material decays. This material was put into a box with a little hole in it to allow a stream of these particles to shoot out. The particles hit a small hole in a lead plate and produced a narrow beam of alpha particles. That was aimed at a thin foil of gold. On the other side of the gold foil was a screen of material that gave off flashes of light when struck with an alpha particle. When the experiment was executed, alpha particles zipped through the lead hole and zipped through the gold as if it were not there.

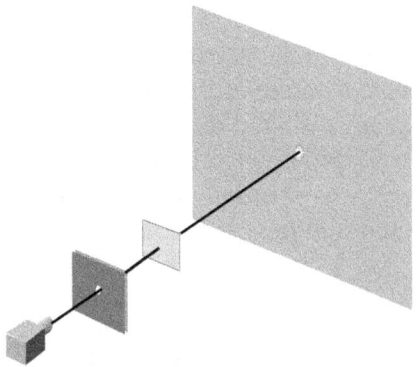

The screen glowed with a spot on it that resembled the size of the hole in the lead plate. At first, Rutherford thought that nothing would happen and the experiment would go nowhere. However, as time passed, flashes were occurring on the screen far away from the bright spot on the screen. The few flashes and the time it took them to appear suggested that there was something very small in the gold that was responsible for the deflections.

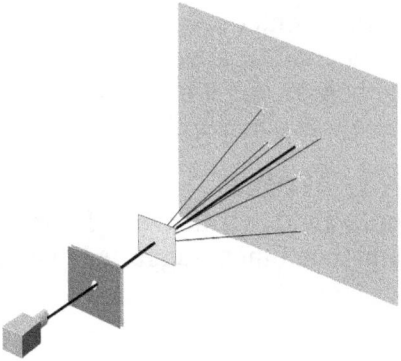

In fact, a few alpha particles were deflected back the way

they came. Rutherford was heard to say, *It was as if you fired a 15-inch shell at a piece of tissue paper and it came back and hit you.* [2]

Counting the flashes on the screen and using probability, they could estimate the size of the thing deflecting the alpha particles as they moved through a gold atom. As the alpha particles were positively charged, the assumption was that the object was positive. The probability they measured indicated that the object was very small. [3]

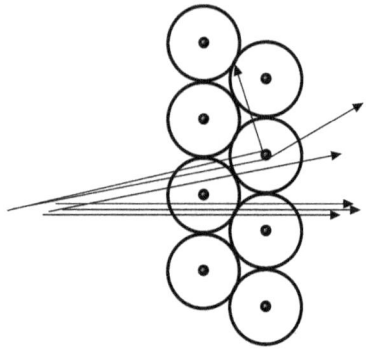

Rutherford announced to the world that the atom contained a positively charged center that was very small and was surrounded by material negatively charged. He, as the careful scientist he was, did not offer any conclusions about what the negative material might be.

The Myth is Born

After Rutherford's announcement and Bohr's observations of hydrogen radiation, others suggested that the electron material around the positive center was like planets orbiting the sun. The following picture represents this idea.

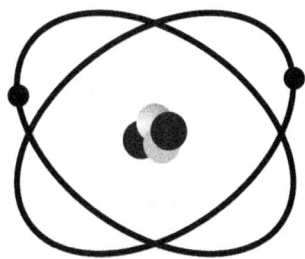

Today, in our popular culture, this is still the model used when the structure of the atom is discussed or drawn in pictures. This model has been demonstrated to be very inaccurate with the study and impact of quantum physics on our society. [4] Yet this model persists from one edge of our society to the other.

Why is this Important?

First, the idea that the atom contained a very hard small middle advanced the understanding of the structure of matter. Then, treating the electron as a little hard ball spinning around the nucleus was very wrong. This triggered a quest to determine why such electrons did not fall into the nucleus. This quest was a major step in developing an understanding of the universe.

Chapter 4

Quantum Mechanics is Born

Max Planck studied the colors of light coming from hot pieces of metal. At the time, the community knew that all bodies gave off electromagnetic radiation or light of varying colors. The natural bumping and crashing of molecules in matter gives rise to these emissions. Likewise, external electromagnetic waves enter bodies and are absorbed constantly. The result is that the body is in equilibrium with its environment. That is, it is giving off as much energy as it is taking in.

Take for example a block of metal. Heat it with a flame or some other heat-producing device. The block heats up and gives off electromagnetic radiation or colors, if you will.

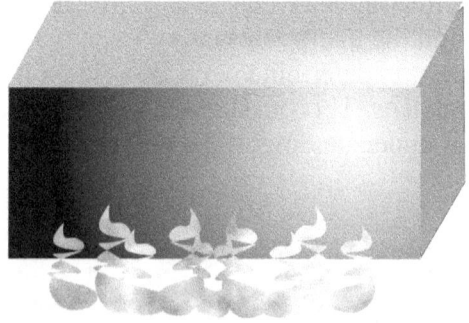

Planck was studying the structure of such bodies to determine how the body radiated electromagnetic energy or light. His theory was that there were little oscillators in the metal. These oscillators absorbed and radiated energy. Here is a rough example of what this means. The spring and ball on the left in the following picture represents an oscillator. When the ball is pulled down and released, it appears as the spring and ball on the right. The ball bounces up and down.

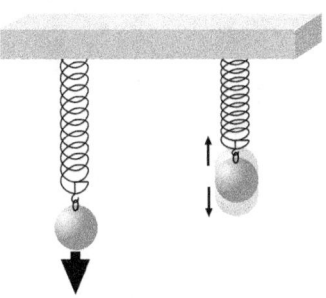

While this is not what goes on inside a piece of metal, it is illustrative. In a sense, the atoms in metal oscillate in a similar way.

At some point, the oscillating ball will emit radiation. That is, the oscillator gives off its energy. Because the energy has left the ball, as depicted in the spring and ball on the right, the ball stops bouncing. The energy given off appears as light.

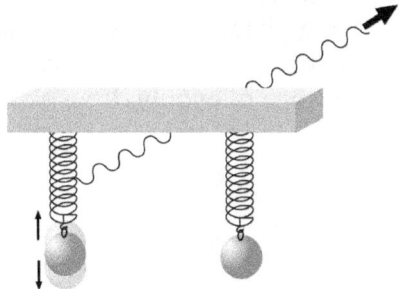

Then the spring and ball stops oscillating, as is shown on the right.

A block of metal contains many of the oscillators that are absorbing energy and radiating it constantly.

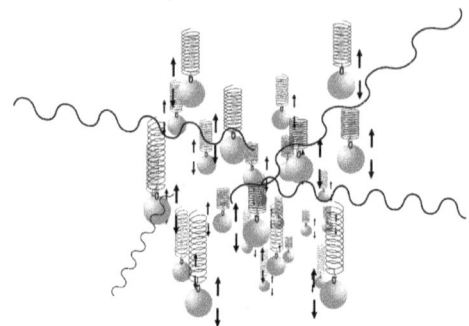

Planck attempted to develop a formula that would predict the energy given off by a number of these oscillators. That is, he attempted to develop a formula that would match the heat actually radiated by a block of hot metal. Planck assumed that the ball could be pulled down at any length and allowed to oscillate at any frequency. This first attempt failed. In exasperation, he attempted other methods. In one method, he based the energy in each oscillator on some integer number of a discrete value. [1] This method produced an equation that worked. That discrete value eventually came to be known as Planck's constant. The following picture depicts that if an oscillator oscillated at some multiple of that value, it could participate in the exchange of energy. If an oscillator attempted to move at some non-integer of Planck's constant, it would not function.

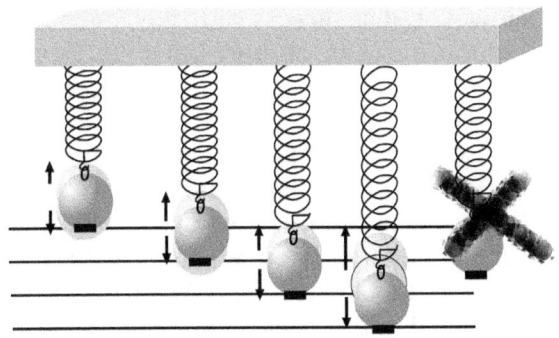

The above picture shows four oscillators bouncing at some multiple of Planck's constant. The mark on the bottom of each ball hits lines that represent multiples of the constant. The spring and ball on the right does not attempt to oscillate at one of those multiples. Therefore, it does not oscillate.

The energy in the radiation from metal is given by Planck's constant times the frequency of the radiation. The equation appears as, $E = hf$, where E is energy, h is Planck's constant, and f is the frequency of the light.

Note that Planck was not interested in the way electromagnetic radiation moved through space. Planck was interested in how the oscillators moved to produce that radiation. Einstein took up the motion through space later.

The important item to carry away here is that the oscillators in the metal only released energy in multiples of Planck's constant. That came to be known as a quantum. [2]

Now Planck's finding did not get a lot of attention when it was first discovered. However, as time passed, Planck's constant appeared in other places. It seemed to answer questions. However, that success generated even more complicated questions. One of those other places was in Einstein's observations of the photoelectric effect.

Note: Planck used the spring model during his study of this phenomenon. We now know there are no springs in metal or atoms. Rather electrons shift from one energy level in the atom to another energy level. When the shift is from a higher energy level to a lower energy level, a photon is emitted as described in this chapter. Energy levels are discussed later in this book.

Why is this Important?
The energies described by Planck's constant became limits in many natural phenomena. These included the wavelength of light and Heisenberg's uncertainty principle.

Chapter 5

Photoelectric Effect

The photoelectric effect occurs when light strikes metal knocking an electron out of the metal. [1] The following picture depicts this.

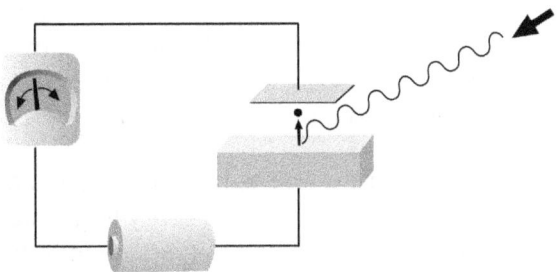

This picture shows an incident beam of light entering the experiment and hitting a block of metal. When the light hits the metal block, an electron is knocked from the block. When the electron moves from the block to the plate, that moving electron is like electricity moving through the wire. This causes the meter to move a bit indicating current is moving through the wire and meter. This process is called the photoelectric effect. As light continuously shines on the block, there is a continuous flow of electricity through the wire and meter.

Einstein Steps In

Einstein proposed that the light incident on metals consisted of small packets of energy.

Electrons could be knocked out of the material if one of these packets had enough energy to break an electron free from the bonds in the material and give the electron some energy to move away from the material. [2]

There are some noteworthy issues here. First, the incident packet of energy must have enough energy to get an electron to break free of its atom. As the amount of energy in a packet of light is proportional to its frequency (based on Planck's formula), the incident packet must have a high enough frequency for the task. This violated Maxwell's theory of light. That is, Maxwell's theory of light predicted that the amount of light determined the effect not the amount of energy in each photon. The observed phenomena and Einstein's theory appeared to oppose the idea that light is a wave. Instead, this theory proposes that light consists of particles of some kind; each having a specific amount of energy given by Planck's constant. In 1915, Robert Andrews Millikan performed experiments that demonstrated Einstein's idea was correct. [3]

The issue to carry away here is that Einstein proposed that

the light hitting the metal and causing the photoelectric effect consisted of localized particles made of quanta of energy. That is, light, in this experiment, was not a wave.

Why is this Important?

This observation contradicted all past observations about the nature of light. This is important for it was a critical step to understand that all aspects of the structure of the universe consist of wave and particle phenomena. The biggest step occurred when de Broglie pointed out that particles are waves like photons.

Chapter 6

The Most Famous Equation

Einstein's

$$E = mc^2$$

is the most famous equation in the world.

He Does It Again

The study of mass and energy equivalence is a long one.
Einstein was not the first to propose a mass-energy
relationship. [1] However, Einstein was the first scientist
to propose the formula. It was the result of his study of
space and time. That was part of his analysis of Special
Relativity.

To get an idea of how comprehensive this equation is,
consider a consequence. If you have block of metal at
room temperature, it will have a given mass. If you heat
the metal so it is hot, the mass of the block will be larger.
The amount is impossible to measure as the change in
mass is so small. However, that is a direct consequence of
Einstein's formula.

Why is this Important?

In the context of our study here, Louis de Broglie used this
formula to suggest that small particles moving through
space are waves. This is a very significant observation in
the emergence of quantum mechanics. We study that next.

Chapter 7

Matter Waves

Quantum mechanics becomes a real science when matter is shown to be a wave phenomenon.

Louis de Broglie

In 1930, a Frenchman named de Broglie looked at the equation of light and made some changes. You are aware that light is considered a wave and has an equation representing its motion through space. The elements of that equation are distance, time, frequency, and wavelength. De Broglie must have observed that light is a wave and a particle as well. Could de Broglie have wondered if a particle like an electron may be a wave as well? If so, a moving electron could have a wavelength and frequency. Consider that Einstein's equation showed a relationship between mass and energy. Also, consider that energy was equal to some number times Planck's constant. The point is that one could use the mass of an electron and convert it to a wavelength and frequency. That is, convert the value of the mass of an electron to energy. Then use that value of energy to determine a wavelength. Then, insert that wavelength into the wave equation for light. The result is that you would get a wave equation for matter. That is what de Broglie did.

In 1927, Davisson and Germer observed an experiment in which electrons moving at a specific speed diffracted with each other. Using de Broglie's matter wave equation, they could measure the wavelength of an electron. [1] As diffraction had been established as a way to demonstrate wave phenomena, the electron was shown to be a wave.

Why is this Important?

De Broglie's observation revealed that light and matter display wave phenomena. It laid a direct path to the development of lasers, LED's, and transistors.

Chapter 8

Uncertainty

The last phenomenon we consider is the Heisenberg uncertainty principle. In brief, it claims that if we attempt to measure the position of a particle, we will get different measurements each time we measure it. Be aware that this only happens when a particle is small such as the size of an electron.

How Does It Work?

Let's do a thought experiment with some pictures to better understand. First, consider something familiar in the universe of the large. The following depicts a rifle shooting a bullet four different times.

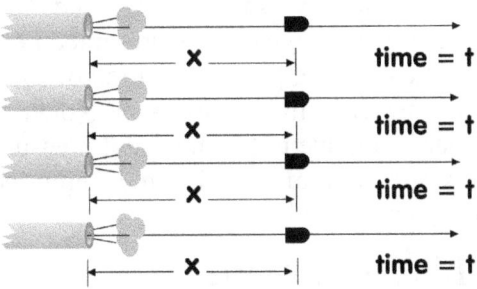

Note that each bullet travels x distance during time t with each shot. That is, if each bullet weights the same, is the same shape, and is shot with the same amount of force; each bullet will travel the same distance in the same amount of time. We are accustomed to this.

This is not true for small particles such as electrons. The following depicts this situation. We assume each electron (the dots in the picture) is shot from an electron gun on the left with the same force. At the end of time t, each electron will have traveled a different distance even though in each try, the time is the same.

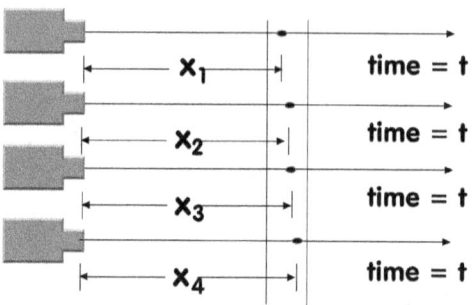

This is quantum phenomena. Note that, although the distances are different, they are relatively close to each other.

Realize that in most texts where this discussed, a lot more is said about it. However, the crux is presented here. When you attempt to measure the precise position of something very small, the answer is always different. Note that the different answer is not due to the means of measurement. The little spots of matter are actually in different positions. This is part of quantum weirdness.

Why is this Important?
We need to consider uncertainty when attempting to understand how electrons move in wires and transistors. This is critical in discussions that follow.

Chapter 9

Pulling it all Together

At this point, we have the critical observations needed for our quest to understand quantum mechanics. Here, we pull the observations together and draw some conclusions. After that, we will be ready to see how electricity flows through wires.

Weirdness is Born

Let's pick up the action just after de Broglie revealed to the world that a wave equation could describe the motion of a particle flying through space. We have seen that a wave equation described the motion of a water wave. You can see the up and down motion of the water and it intuitively matches the form of the equation. With the understanding that light is a wave, we can intuitively see that the equation represents light as it moves through space. However, when we think of a particle moving through space, we have a problem trying to visualize a small hard something fitting into the equation. The question is, what does the de Broglie matter wave equation represent? The equation makes a flying electron look like a water wave and a light wave.

What happened at this point created a mass of confusion about quantum mechanics. Louis de Broglie, at the time, was a young man attempting to get a doctor's degree in physics. His thesis paper presented his theory of matter waves. The professors assigned to judge the worthiness of de Broglie's paper did not understand the meaning of the work. A copy was sent to Einstein for his opinion. Einstein responded that Louis de Broglie should get a doctor degree. [1] However, Louis de Broglie had little to say about how his invention was to be interpreted. Max Born along with Werner Heisenberg decided that the equation represented the probability of where a little hardball was at any particular point in time. [2] Some in the science community agreed and some did not agree. A squabble ensued and left the public confused about what quantum mechanics was about.

One of the supporters of the wave probability approach was Niels Bohr. His notions of reality and consciousness ultimately shaped how the public saw quantum mechanics. He was one of those people that believed the electric flashes inside one's head were some kind of representation of the real world outside the head. That is, he had the point of view that the flashes were our interpretation of reality. Consequently, he had the idea that human observation somehow had power to shape reality. Here are some of his quotes about his point of view. [3]

"It is wrong to think that the task of physics is to find out how Nature is. Physics concerns what we say about Nature."

"A physicist is just an atom's way of looking at itself."

"Everything we call real is made of things that cannot be regarded as real."

At the time, Niels Bohr was a power figure in physics. He had a long list of accomplishments behind him and he ran the most noteworthy institution in theoretical physics. [4] Bohr believed that when it came to quantum mechanics, no picture and no model of any kind would be useful in describing quantum mechanics. He believed that any description of the phenomenon must be mathematical.

Bohr's attitude was revealed in a meeting where Richard Feynman presented his work on Quantum

Electrodynamics. When Feynman drew his pictures of photon and electron interactions on a blackboard, Niels Bohr leaped out of his seat screaming indignation. [5]

Einstein's response to Bohr's arguments was interesting. Bohr published his own thoughts about his and Einstein's conversations on the subject. Einstein's opinions were not published. In a private correspondence, Einstein said Bohr was a, "talmudic philosopher [who] doesn't give a hoot for 'reality,' which he regards as a hobgoblin of the naive..." [6]

Without going into all of the philosophy, here is the upshot of it all. De Broglie and Einstein thought there was more to the matter wave equation. They thought the wave equation was a mathematical representation of something deeper. [7] [8] That is, they thought there was something physical about it rather than some human interpretation of observed probabilities. Eventually, Schrödinger came to believe the same. [9] These three greats pursued this line of thought. Neils Bohr did not. Therefore, the probability aspect ruled.

However, the probability approach worked. There is no logical connection between an electron flying through space and the matter wave equation being a probability curve. Yet, when applied to electrons moving around a nucleus in an atom, the resulting numbers are very accurate. The equations yield answers to the behavior of electrons moving around the atom. Due to this, many believe the math describes what the electron is. As the equation matches the math of a water wave, these many believe the electron is a wave of matter of some kind. This is as if saying ten grams describes a ten-gram apple.

In essence, we have no idea what an electron is. We know that when we have any given number of electrons near a nucleus, we can map out the space where they probably exist. Is it a wave? If so, what is the electron a wave of? Is it a little hard ball? If so, how can it be in several places, as it would need to be in an atom.

The result, presented in this book, will take Born's, Heisenberg's, and Bohr's thoughts of reality and show how they have become what we call quantum mechanics.

Here is What Happened

Max Born blended Heisenberg's uncertainty principle with de Broglie matter wave function to describe the probabilistic behavior of small particles. While the equation made sense when talking about water and light, what did it mean for matter? The first question is, "What could the variable y represent in the following equation?"

$$y = A \sin \left(2\pi \frac{x}{\lambda} - 2\pi \frac{t}{T} \right)$$

Then, what does the wiggly curve represent? The curve has come to represent the probability amplitude of the existence of the particle as it moves from the left to the right. The y in the above equation is replaced with the Greek symbol Psi.

$$\psi = A \sin \left(2\pi \frac{x}{\lambda} - 2\pi \frac{t}{T} \right)$$

You may wonder why the Greek symbol came to be in quantum physics. That symbol is commonly used in probability and statistics. If you look up PSI on the internet, you will find it is an acronym for Probability, Statistics, and Information.

The probability of existence of the particle at any point is then the square of the probability amplitude. That is:

$$\text{Probability} = \psi^2$$

Now, consider an experiment where a small particle such as an electron is shot out of a gun.

If we believe the electron is moving according the equations shown, this picture changes. Then, the electron seems to oscillate up and down as it moves across the page.

Note, however, that the equation is squared. The most notable consequence of doing this is that the negative parts of the matter wave become positive. When we square Psi, the curve appears as follows.

This curve represents the probability of where the particle could be as it moves to the right. The dots under the curve represent where a particle would be during this trip. At first glance, it does not make sense. In this scenario, there is only one particle. Why does it seem it can be at all places? This is not the point of the curve. The curve represents where the particle can be as a function of x. Let us look at part of the curve closer.

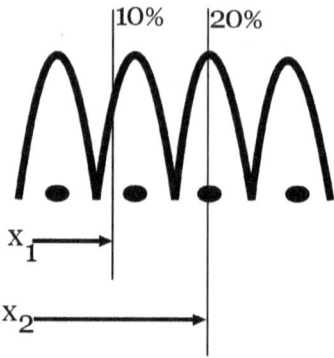

Here the wave function implies that if the position of the particle is supposed to be at x sub 1, the probability of the particle being there is 10%, *when the value of x in the wave function is x sub 1.* The probability of the particle at x sub 2 is 20%, *that is when the value of x in the wave function is x sub 2.*

Bear in mind, the curve does not represent the particle moving to the right. It is a representation of the position probability as the particle moves to the right.

Let us attempt to clarify this a bit more.

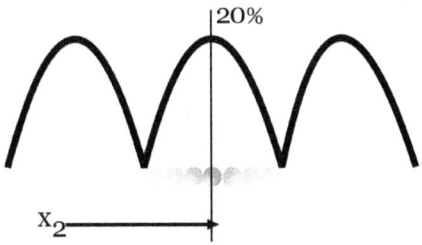

When the variable x in the equation is x sub 2: the probability of the particle being there is 20%. However, as the picture attempts to depict, the particle may be farther than that or less than that.

Again, this does not make sense in a common sense way. This is observed in the world of the very small. This is quantum physics. If you try to follow a logical path of thought about why this happens, you will have a great deal of difficulty. This path was developed through trial and error. At some point, the concepts were tried with real world experiments. The experiments produced real world answers. Therefore, *It works!*

Observing the Flow of Many Particles

Often when this subject is discussed, one does not realize that we are observing many events. In this situation, we should observe many electrons being shot from the electron gun. If millions of electrons are coming out of the gun, the effect of probability suddenly becomes very apparent. The following represents the matter wave functions just presented.

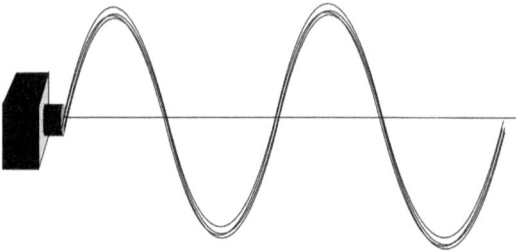

The following represents the flow of very small particles emitted from the gun.

Note that where the probability amplitude has a large value, the beam is darker indicating more particles are there. The light areas represent a smaller number of particles present. Again, keep in mind that this is the result of shooting many, many particles. With many such events, the probability of appearance increases so much that the combined probabilistic occurrences seem to become solid. That is, if a million electrons shoot out of the gun each second, when you put your finger in the middle of the black area, you will find an electron. The interesting thing is that when you put your finger in the light area, you will not find an electron. Not only that, the electrons will appear to be to the left and right of your finger. It appears that the electrons, assuming they are traveling from left to right, appear to travel through your finger or suddenly disappear from the left of your finger and appear to the right of your finger. This is quantum weirdness in action.

Consider the following picture that depicts the center of a particle moving through space. This is a picture of where the center of the particle should be at three different points in time.

From the foregoing, we understand that the center might be in front of those positions or behind them. The following picture depicts this. That is, because of the motion of all the photons in the particle, the center or average position of the particle may be in front of or behind the expected center of points. This picture attempts to show a variety of forward and backward positions.

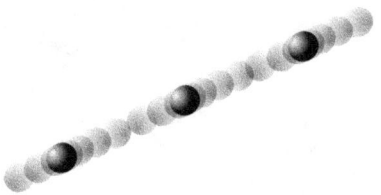

Of course, the particles may exist higher or lower.

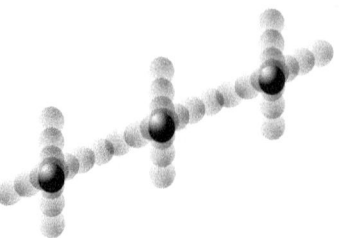

Moreover, the probability of existence of the particles may be side to side.

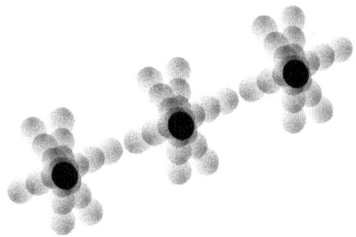

If we draw a line representing the probability of where the center might be, the lines would appear something as follows.

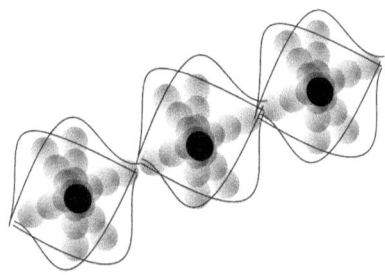

The result is that if we saw a number of electrons flying

along side by side, they would appear as an oscillating cloud moving through space.

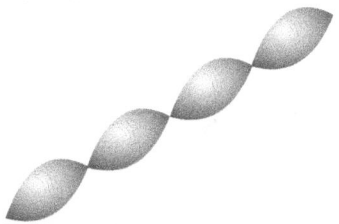

Now, use some caution with this interpretation. The interpretation might be that the electron is a small ball that has this weird behavior. A quantum physicist will jump up and down telling you that is wrong. They will claim that the motion of the electron is described with a mathematical probability curve. There may not be little hard balls there. There may be nothing there. We just don't know. However, the math tells you what might happen in that space when encountering other, "probability clouds." The best we can do is walk away with the idea that when working with electrons they seem to move as cloud waves. They do not appear to act as little hard balls flying through space.

Why is this Important?

All of this should convice you that the electron is not a little hard ball flying through space. Instead, we should use wave mathematics to plot where an electron might be. Inside of an atom, the electron has a chance to spin around many, many times. Thus, the probability increases at any location to the point, something is there. Treating electrons as electron waves enables atom structure to be, not only explained, but also used.

Chapter 10

Energy Bands

In this chapter, we continue the study of how electrons flow around the nucleus.

Electrons as Planets

The primary issue here is the way electrons form energy bands or levels. To clarify some issues, let's begin with a discussion of the initial way electrons were thought to go around inside an atom after Rutherford discovered the nucleus. The following picture shows electrons as small balls that spin around the nucleus like planets spinning around the sun.

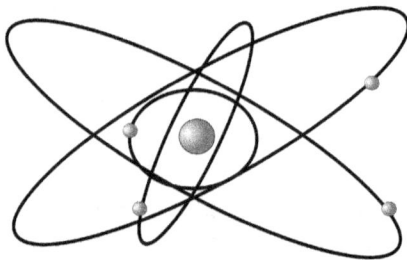

With this image, one would assume that the little balls could spin around with a variety of energies, speeds and at various distances from the nucleus. When one observes the actual behavior of the electrons, this assumption appears false. Let's pursue this notion a bit more to demonstrate its inaccuracy.

Back then, further study revealed the following. The electrons appeared to exist at distinct levels around the atom. Furthermore, precise numbers of electrons existed at each level. The following shows a hypothetical atom with many electrons around the nucleus.

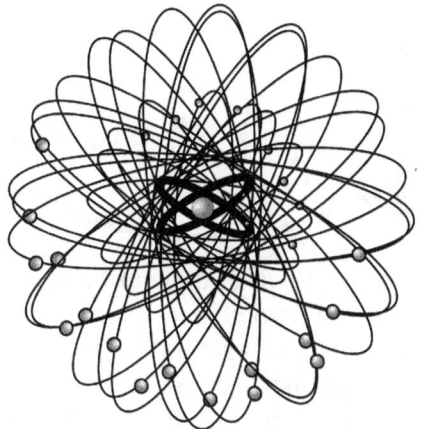

Two electrons appear at the same distance or lowest energy level in the atom. These are displayed as the two thick black ellipses in the middle of the above picture. Next up, eight electrons would be found all at the same distance or one-step higher energy level from the nucleus. Above this, there would be a third level with 18 electrons. There was no reason for the little balls to spin around at those specific levels with those specific numbers. This picture does not make sense. There are other issues with this scheme. However, we need not go into them here.

Electron Waves?

The de Broglie matter wave equation opened a door to an explanation of why the electrons behaved in this odd way. Consider the two lowest electrons near the nucleus. If each electron were some kind of a wave, there would be enough distance around the nucleus for two waves to exist. Here is a representation of two waves not wrapped around a nucleus. One is up and the other is down. You can picture each of these waves like two lengths of string tied to poles and vibrating up and down.

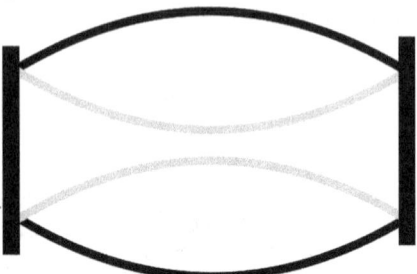

We need to invoke the Pauli Exclusion Principle. In short, this principle states that two objects cannot occupy the same space at the same time. Near the nucleus, there would only be enough distance around the nucleus to support the length of one-half wave. Here is a two dimensional representation of a single half wave oscillating about a nucleus.

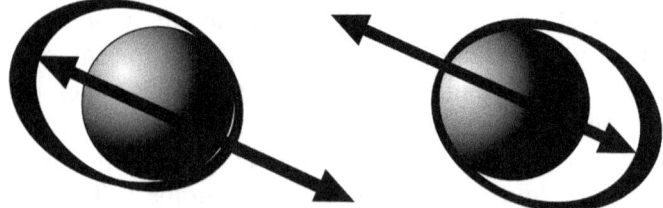

Two electron waves can exist there because both waves are oscillating in opposite directions from each other. While one wave is going left, the other is going right. Therefore, they are alternating positions in the same space.

Each electron would appear like a shell around the
nucleus. Each would be oscillating back and forth. Since
each is on opposite sides of the nucleus, they never occupy
the same space. Since there is no room for another wave in
the distance at the lower energy level, no other electrons
can exist there.

One level up, the scenario is different. There is space for
two half waves to form. Consider how many different
waves can exist in the distance of two half waves.

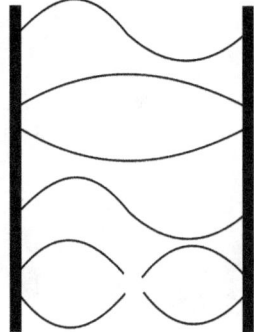

Thus, one level up, eight waves can exist. Here is an
image of an atom with ten electrons around it. It is neon.

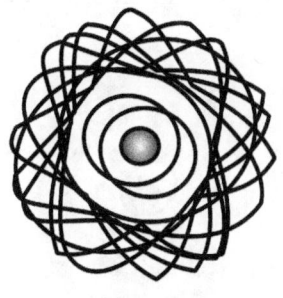

Neon

The point is this. If the electron is interpreted as a matter wave, we can see why there would only be two at the lowest level, eight at the next level, eighteen at the next level, and so on.

Levels and Bands

In essence, we get energy levels or energy bands around the nucleus. The following depicts empty energy levels.

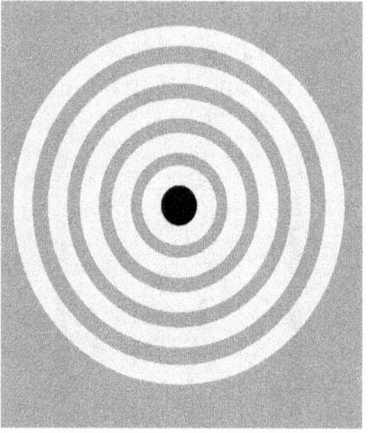

Lets put in one electron wave as would occur with hydrogen.

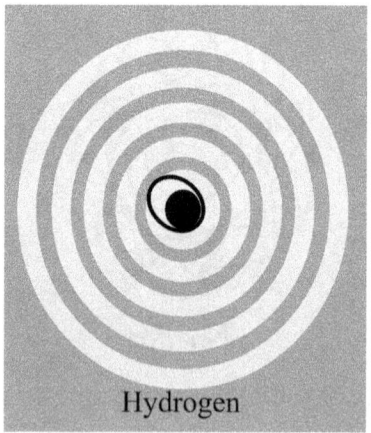

Hydrogen

Here is a diagram showing the electron structure of copper.

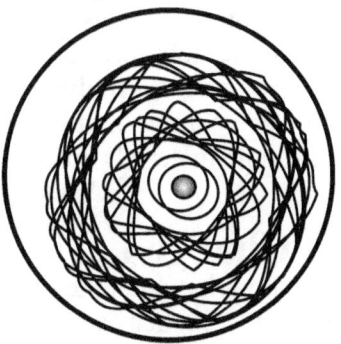

The outermost wave represents a single electron in the outermost valence band. A valence band is a level in which the electron waves are at their lowest level.

Electrons Can Change levels

An important aspect of the level concept is that electron waves can change levels. Normally in solids, the atoms are vibrating due to heat in the solid. This vibration can be transferred to an electron wave, giving it more energy. Then the electron can jump to a higher energy level. Here is a picture denoting this.

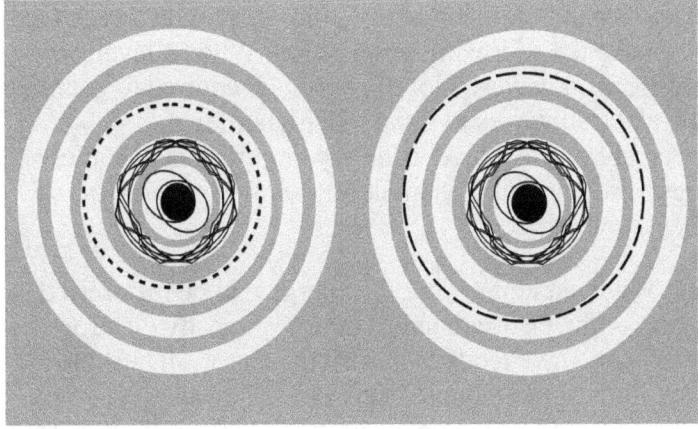

In the above picture, the diagram on the left shows an atom with the outermost electron wave at its lowest energy level. In the diagram on the right, the outermost electron wave is shown having jumped to a higher energy band. As the atoms in solids are constantly vibrating due to heat, the outermost electron in some materials are always in the higher level.

Some Cannot Change Levels

Now let's consider an atom in which the outer electron waves tend not to jump to a higher band. A good example is sulfur. Its electron structure appears as follows.

Note that the outermost level of this atom has six electron waves. When this atom is shaken, electrons in the outermost band will tend not to jump to a higher level. One could make the following observation. When a level has many electron waves, a lot of energy is required to get one of the electron waves to jump to a higher level. Conversely, when a level is nearly empty, somewhat less energy is required to get an electron wave to jump to a higher level.

Why? When an atom is hit, the energy can be transferred into one of the energy bands. If there are many electron waves, each wave tends to absorb the energy. Thus, the energy hit is split amongst many electron waves. The energy devoted to a single electron is reduced. If however, there is only one electron wave in an energy band, all of the hit energy is taken in by that single electron. Subsequently, that electron is more likely to bounce out of its lowest energy band and jump to a higher band.

Consider this analogy. Put nine marbles in a glass and hit the glass on the bottom. The marbles will jump but not high. Now, put one marble in the glass then hit it with the same amount of force.

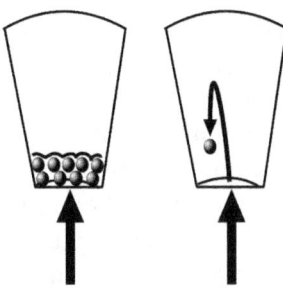

The single marble jumps higher than the bunch of marbles.

You might ask, "Where does the energy go if an electron does not jump energy bands?" Well, when an electron does not jump bands, the atom as a whole moves depending on the energy that hit it.

Why is this Important?
When electrons do not jump to other energy bands, we have a non-conductor. When electrons jump to other energy bands, we have a conductor. This forms the basis of how a transistor works.

Chapter 11

How a Wire Works

We are now in a position to understand how electricity moves through a wire. Many will find this a curious next step for it appears mundane. While many explanations of quantum mechanics say that quantum theory explains how transistors work, they do not continue on to explain that. Often they extol the virtues of QM claiming that consciousness is the cornerstone of existence. However, the next mundane but important step in a discussion of quantum mechanics is to show how electricity moves through a wire. It is a critical next step. It follows from a discussion of quantum energy levels.

Wire Atomic Structure

We have already looked at a copper atom and had a peek at the outer electron wave moving to a higher energy level. That is the main ingredient in the conduction of electricity in a copper wire. Let's visualize the construction of a wire by placing a number of copper atoms side by side.

In this picture, we have placed four copper atoms side by side. Each atom has three main parts to it. The center holds the nucleus and the lower energy levels. These are shown together as the center disk. Moving outward from the center of each atom, we have the outermost valence layer or band. The valence band refers to a band that contains the electrons for the atom when all electron waves are at their lowest level. In this picture, each outermost valence band contains one electron wave depicted by the dashed lines. That line represents the electron wave oscillating in each outermost valence band. Again, moving outward from the center of each of four atoms is another band called the conduction band. That is the energy band that conducts electron waves through the wire. Notice, in the above picture, that the conduction bands overlap.

The following picture depicts each electron has jumped to a higher level.

Now each of four electron waves is in the conduction energy band. Notice that each electron wave is also in its neighbor's conduction band.

The following picture illustrates the space created when the conduction bands of all copper atoms touch. The conduction band forms a consecutive area an electron wave can roam through.

The following picture depicts one electron wave moving through this conduction space.

Electricity flowing through a wire then consists of the space created by the conduction bands touching.

Electron Motion

In the following picture, we have attached a battery and some wires to the string of copper atoms. The negative end of the battery pushes electrons along a wire into our four sample copper atoms. The electrons build up at one end of the chain of four. The electrons gather on the left and push into the conduction band. This forces the electrons in the conduction band to move to the right. The path leading to the other end of the battery has fewer electrons there and the electrons from the conduction band move toward the battery.

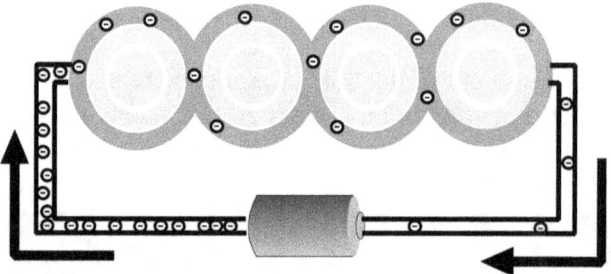

Non-Conduction

To be complete, let's look at non-conductors and semi-conductors. The following picture represents a non-conductor. In this representation, the outermost valence band contains eight electrons such as sulfur. Here a lot of energy is required to cause the outermost valence electrons to jump to a higher level. Thus, they commonly do not jump there. As there are no electrons in the conduction band, electricity does not flow through the chain.

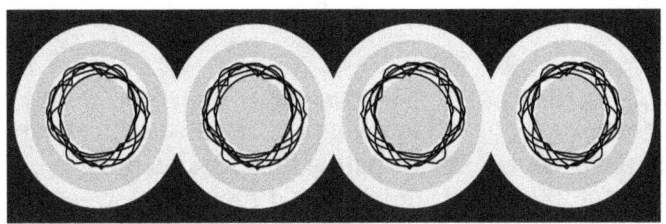

Semi-conductors are halfway between conductors and non-conductors. There are several electrons in the outermost valence band. Heat vibration will cause the outermost electrons to jump into the conduction area. However, the electrons do not move as freely as electrons in a conductor.

Why is this Important?
Understanding the way electrons move through a wire prepares us for a major step. That is, how does electricity move or not move through doped semi-conductors.

Chapter 12

Diodes

Right now, we are on a path to see how transistors work.
Understanding how a diode works is the most critical part
of that path. We began this path with a look at energy
levels. After that, we examined the difference between
conduction, semi-conduction, and non-conduction. The
next step is to understand how mixing different
semi-conductors together can produce materials that
enable electron waves to move or not move through those
materials. The process of mixing is called doping. We will
examine mixing a small amount of arsenic with pure
silicon and a small amount of gallium with pure silicon.

Doping Silicon with Gallium

First, consider doping pure silicon with gallium. Gallium
has three electron waves in its outside valence band.

**Gallium
3 electrons in
outer band**

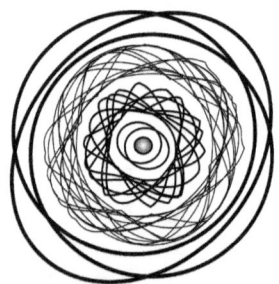

As mentioned in the chapter on electrons moving through
wires, the electron waves move randomly in the
conduction bands. Here, this is a mix of silicon and
gallium. Because gallium only contributes three electrons
to this random motion, occasionally there will be only

three electrons near a silicon atom. During that moment, the vicinity of the silicon atom has a plus charge of one. That plus charge will exert a force on other electrons to move toward that silicon atom. In descriptions of how transistors work, the plus one charge is referred to as a hole into which an electron could fall.

A common point of confusion in most materials presented on this subject is the statement that the gallium only has three outer electrons that create a hole in the material into which an electron can fall. While this is a true statement, how this is achieved is not clear.

The following diagrams are designed to explain this phenomenon. In the following picture, four electron waves are depicted moving through the conductive bands in silicon that has been doped with gallium. In this depiction, the four electrons happen to pass by a silicon atom.

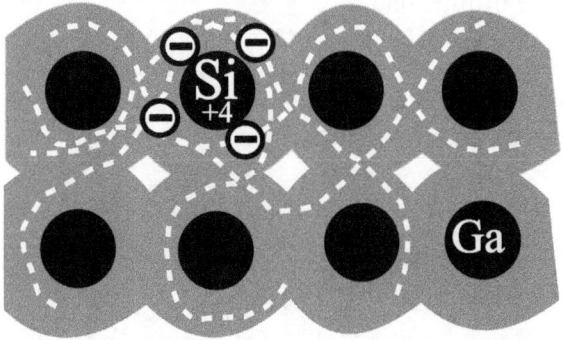

The charge on the silicon atom is plus four. As there are four electrons near the silicon atom, each contributes a minus one adding up to a minus four. This combines with the plus four to equal a charge of zero. In the area around that silicon atom, at that moment, there is no attraction or repulsion. The charge is balanced or zero.

As time advances and the electron waves move about

randomly, there will be a moment when only three electron waves are in the conduction band of the silicon atom. The following picture depicts this.

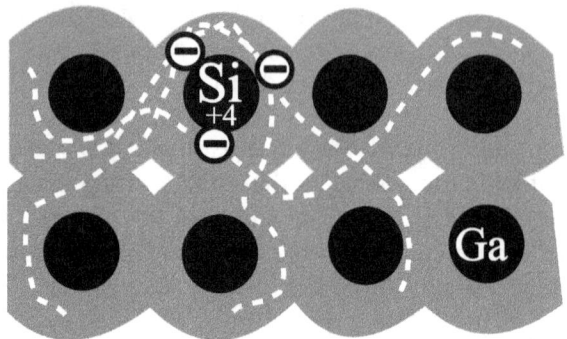

As there are three minus charge electrons near the silicon atom, and the silicon atom has a charge of plus four, the net charge is plus one. Thus, for a moment, that silicon atom is considered a hole for an electron to fall into for it has a charge of plus one.

Because the electrons in the conduction band are constantly moving, the hole can appear somewhere else. The following picture depicts this.

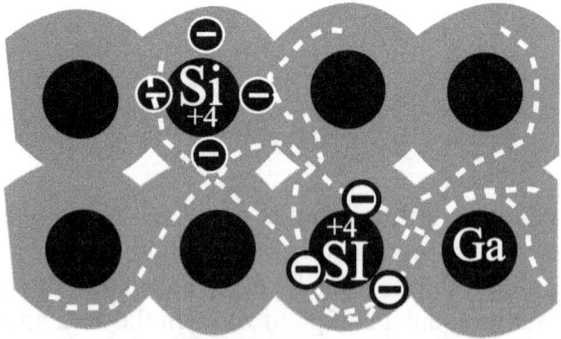

In the location we were considering, the picture shows that four electrons have moved into that space. The random motion of the moving electron waves has produced a plus one situation on a different silicon atom. In the lower right location, only three electron waves are there. Thus, in the space around that atom the electric charge is plus one.

Keep in mind that this effect occurs because there is a gallium atom present. It creates the effect that there is one less electron in the mix of electrons moving about in the block of doped silicon.

Note that these holes appear throughout the material. Note that the holes appear to be moving randomly around. Of course, the holes are not moving. Electron waves are moving about and away from some silicon atoms creating the illusion the holes move. Because there are many atoms of gallium, there will be many of these holes. Also, note that the charge on the whole block is zero.

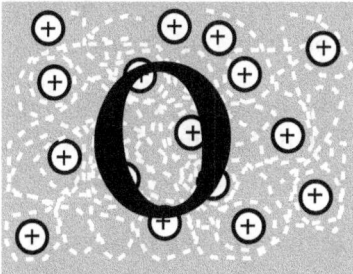

The holes of plus one charge are limited to a very small areas. The block is referred to as p-type for a positively oriented doped silicon block.

Doping Silicon with Arsenic
Next, let's look at mixing a small amount of arsenic with

silicon. The electron wave map of arsenic appears as follows.

**Arsenic
5 electrons in
outer band**

The interesting thing about arsenic is that it has five electron waves in its outermost valence energy level. That is, when all electron waves are at their lowest energy level, the outmost band has five electrons. When the arsenic atom is subjected to normal heat vibration, the five electron waves can enter the conduction band and enable electron wave motion though the arsenic material.

In doping silicon with arsenic, a small amount of arsenic is mixed with silicon. This has the effect of injecting extra electron waves into the conduction band.

In the example with gallium, the random motion of electrons produced a plus charge around silicon atoms. With arsenic, due to it having one more electron in its outermost level than silicon, some number of silicon atoms will have a charge of minus one. The picture below shows five electron waves randomly gathering around a silicon atom. At that moment, the electric charge is negative one.

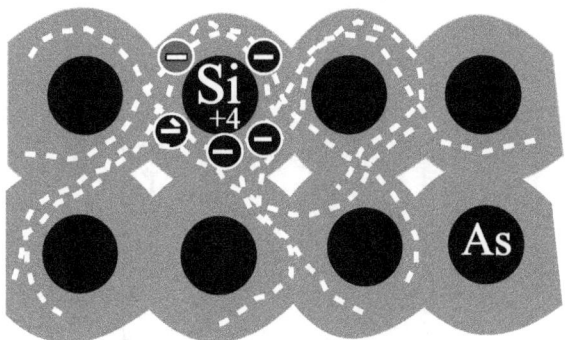

As before, the charges are local to silicon atoms. The block as a whole has no or zero charge.

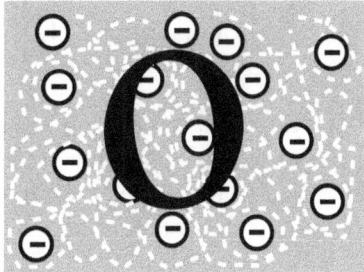

The arsenic doped silicon is referred to as n-type. That is, "negative type." Both blocks conduct electricity like any other semi-conductor.

Building a Diode

A diode is formed when we put a block of gallium doped

silicon next to a block of arsenic doped silicon. Here is a picture of that arrangement.

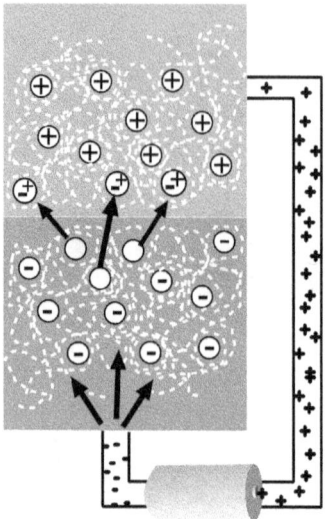

The blocks have been attached to each other and a battery has been attached to the blocks. The plus side of the battery has been attached to the upper p-type block and the negative side of the battery has been attached to the lower n-type block.

Note the battery is trying to push electrons into the n-type block at the bottom. As there is a build up of negative charge in the lower block, electrons are pushed out the top of the block into the p-type block. This is shown by the minus signs beside some plus signs in the upper block.

Eventually, enough electrons flow into the upper block so all the positive silicon atoms can be neutral. Now, the entire upper block is negative because it has many more electrons in it than before.

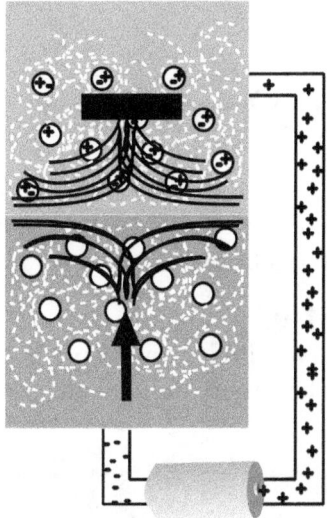

The negative charge on the upper block prevents any more electrons from moving from the lower block to the upper block. The electrons stop flowing through the diode. The pn diode has done its job.

If the battery is reversed so it forces electrons into the p-type upper block, electrons will flow through the diode.

Note that the pressure from the battery caused the barrier at the junction of the p and n type blocks. Note that no current flows. The voltage of the battery keeps the barrier there.

Why is this important?
This describes the heart of a diode's behavior. When a voltage falls across the junction of the p-type doped silicon and n-type doped silicon, a barrier is formed through which electrons cannot pass. Our next step is to see how this enables a transistor to function.

Chapter 13

Making a Transistor

Now, we have all the ingredients to make a transistor. We just plug one into the other.

One Small Step

A transistor is like two diodes on top of each other. Let's begin with a diode that is conducting electricity.

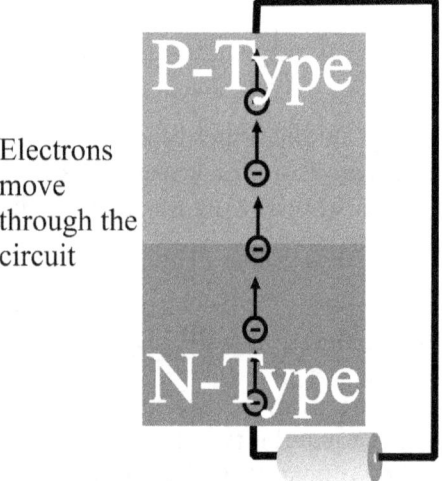

Electrons move through the circuit

The positive side of the battery is attached to the p-type part of the diode. The negative side of the battery is attached to the n-type part of the diode. The electrons move through the circuit.

Here is a diode that is not conducting.

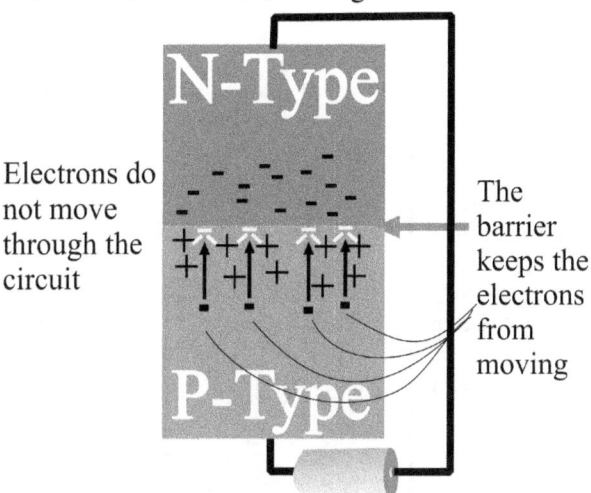

Electrons do not move through the circuit

The barrier keeps the electrons from moving

The positive side of the battery is attached to the n-type part of the diode. The negative side of the battery is attached to the p-type part of the diode. The electrons do not move through the circuit. As mentioned, the voltage pressure from the battery keeps the barrier in place preventing the electrons from crossing the barrier.

Now, let's put these two diodes together and see what happens.

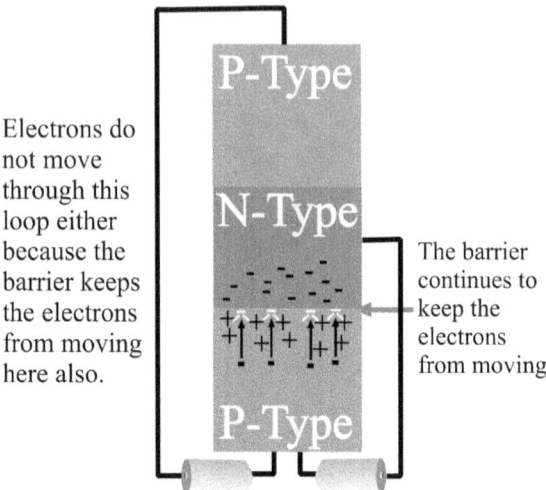

We still have two circuits but both are attached to the stacked diodes. The circuit on the right still keeps the barrier in place due to the voltage pressure from the battery. Electrons do not move in the circuit on the left due to the barrier.

Consider what happens when we break the circuit on the right.

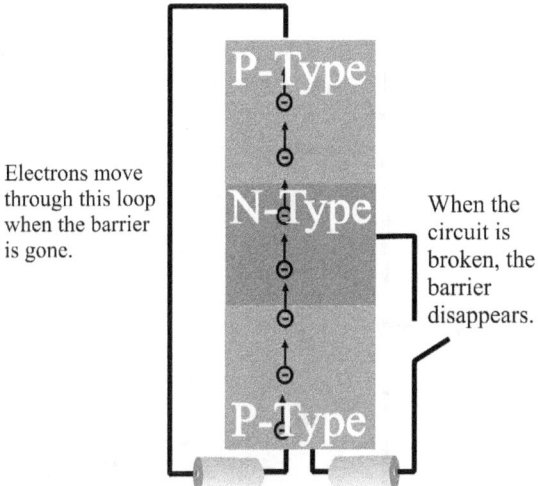

Electrons move through this loop when the barrier is gone.

When the circuit is broken, the barrier disappears.

When the circuit is broken, the voltage force from the battery is no longer pressing on the electrons in the barrier. With the barrier gone, the electrons move through the circuit on the left.

From this, you can see that opening and closing the right circuit controls the flow of electrons through the circuit on the left. Note, the amount of voltage in the right battery need not be large to control the electrons moving through the circuit on the left. Thus, the transistor appears to amplify the signal (voltage) from the battery on the right.

A Sound System
In a sound amplification system, something is placed in the circuit break that causes the voltage to change. Let's put a small can of carbon particles. That represents a microphone. When we talk into the microphone, the carbon particles shake depending on the changing tone of our voice. The resistance of the particles change

depending on how much they are shaken. This varies the
voltage going into the circuit on the right. In turn, this
affects the strength of the barrier. As the barrier changes,
the amount of electron flow in the circuit changes. When
that wire is run through a speaker, our voice appears to be
amplified in the speaker.

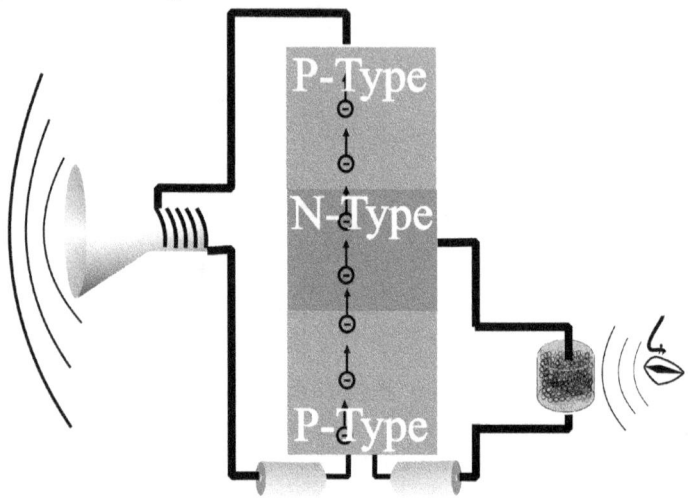

Chapter 14

Conclusion

Perhaps the most important concept to carry away after you read the last page of this book is that an electron is not a little hardball flying through space. It is important to know that we do not know what it is. However, we have mathematics that describes where it can be now and then. This is somewhat like saying 80% of the people like cherry pop. Essentially this lets cherry pop makers know that if they keep making it they will keep selling it. The 80% number tells you nothing about what cherry pop tastes like or what the people are like that drink it.

The mathematics suggests that electrons are shaped somewhat like clouds. In solids, the clouds touch each other. Here is a visualization of that.

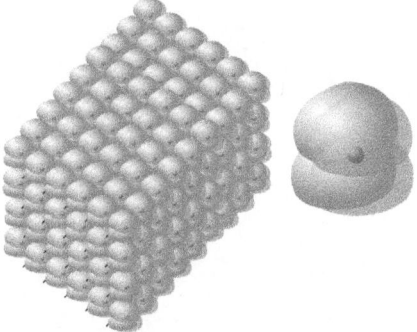

The figure on the right is an attempt to represent electrons surrounding a nucleus. The figure on the left is a group of these atoms forming a solid. Note that the nucleus has been found to be another kind of cloud. It is perhaps more dense. Here is the crux of quantum mechanics. The world does not consist of little hardball things. The world of the small consists of some kind of wave things that can bounce off each other or stick to each other. The deciding factor of how an atom reacts with other atoms depends upon how many protons are in the nucleus and the idea that electrons act like waves in the atom.

The details of water, light, and matter waves work together with a common wave equation to represent the structure of the universe. All of this is supported with the Heisenberg uncertainty principle, Planck's constant, Maxwell's equations, Einstein's observations, and the myriad of experiments done by scientists during the last 2000 years.

References

Foreword
No References

Introduction
No References

Chapter 1: Water and Light Waves
1. scienceworld.wolfram.com. Ether. Retrieved June 5, 2011 from http://scienceworld.wolfram.com/physics/Ether.html

2. britannica.com. Huygens   principle. Retrieved October 12, 2011 from http:// www.britannica.com/EBchecked/topic/ 277804/Huygens-principle

3. en.wikipedia.org. Corpuscular theory of light. Retrieved October 13, 2011 from http://en.wikipedia.org/wiki/Corpuscular_theory_of_light

4. en.wikipedia.org. Diffraction. Retrieved October 13, 2011 from http://en.wikipedia.org/wiki/Diffraction

5. en.wikipedia.org. Diffraction formalism Retrieved October13, 2011 from http:// en.wikipedia.org/wiki/Diffraction_ formalism

6. sci.tech-archive.net. Theories of light 2: ether theories.

Retrieved October 13, 2011 from http://sci.tech-archive. net/Archive/ sci.physics/2005-10/msg01264.html

7. en.wikipedia.org/wiki. Double-slit experiment. Retrieved October 13, 2011 from http://en.wikipedia.org/wiki/Double- slit_experiment

8. en.wikipedia.org. Huygens-Fresnel principle. Retrieved October 12, 2011 from http://en.wikipedia.org/wiki/ Huygens%E2%80%93Fresnel_principle

Chapter 2: Maxwell's Equations
1. en.wikipedia.org. A Dynamical Theory of the Electromagnetic Field. Retrieved October 28, 2011 from http://en.wikipedia.org/wiki/A_Dynamical_Theory _of_the_Electromagnetic_Field

2. en.wikipedia.org. Wave Retrieved October 28, 2011 fromhttp://en.wikipedia.org/wiki/Wave

Chapter 3: The Biggest Scientific Myth
1. galileo.phys.virginia.edu. Rutherford_Scattering. Retreived October 6, 2011 from http://galileo.phys.virginia. edu/ classes/252/Rutherford_Scattering/ Rutherford_Scattering.html

2. en.wikipedia.org. Geiger-Marsden experiment. Retrieved October 6, 2011 from http://en.wikipedia.org/wiki/ Geiger%e2%80%93Marsden_experiment

3. hyperphysics.phy-astr.gsu.edu. Rutherford Scattering. Retrieved October 6, 2011 from http://hyperphysics.phy-astr.gsu.edu/hbase/rutsca.html

4. en.wikipedia.org. Rutherford model. Retrieved October 6, 2011 from http://en.wikipedia.org/wiki/Rutherford_model

References 75

Chapter 4: Quantum Mechanics is Born
1. encyclopedia.com. Max_Planck. Retrieved October 5, 2011 from http://www.encyclopedia.com/topic/Max_Planck.aspx

2. abyss.uoregon.edu. Planck s constant. Retrieved October 4, 2011 from http:// abyss.uoregon.edu/%7Ejs/21st_ century_science/ lectures/lec12.html

Chapter 5: Photoelectric Effect
1. newworldencyclopedia.org. Photoelectric effect. Retrieved October 4, 2011 from http://www.newworld encyclopedia.org/ entry/Photoelectric_effect

2. spiff.rit.edu. Einstein and the photoelectric effect. Retrieved October 4, 2011 from http://spiff.rit.edu/classes/phys314/ lectures/photoe/photoe.html

3. www.britannica.com. Robert Andrews Millikan. Retrieved October 4, 2011 from http://www.britannica.com/EBchecked/ topic/382902/Robert-Andrews-Millikan

Chapter 6: The Most Famous Equation
1. Miller, Arthur I. (1981), Albert Einstein's special theory of relativity. Emergence (1905) and early interpretation (1905 1911), Reading: Addison Wesley, ISBN 0-201-04679-2

Chaapter 7: Matter Waves
1. britannica.com. Lester Halbert Germer. Retrieved September 30, 2011 from http://www.britannica.com/EBchecked/ topic/231764/Lester-Halbert-Germer

Chapter 8: Uncertainty
No References

Chapter 9: Pulling It All Together

1. http://www.nndb.com/people/313/000072097/

http://www.netplaces.com/einstein/
einsteins-contemporaries/ louis-de-broglie-18921987.htm

2. en.wikipedia.org. Probability amplitude. Retrieved
October 2, 2011 from http://en.wikipedia.org/wiki/
Probability_Amplitude

3. http://physics.about.com/od/nielsbohr/tp/
Niels-Bohr-Quotes.htm

4. Aaserud, Finn. "History of the institute: The
establishment of an institute". Niels Bohr Institute.
Archived from the original on 5 April 2008. Retrieved 11
May 2008.aaa (Einstein to Schroedinger, June 19, 1935.
Translation from Don Howard, "Einstein on Locality and
Separability," in Studies in History and Philosophy of
Science, 16 (1985), pp. 171-201 on p. 178.)

5. The Illusion of Reality. BBC Atom Part 3 of 3 Southern
Star Entertainment UK Pic MMVII A video presentation.
27:00. Retrieved June 8, 2011 from http://www.youtube.
com/watch?v=bF5-jTlolMk

6, Einstein to Schroedinger, June 19, 1935. Translation
from Don Howard, "Einstein on Locality and
Separability," in Studies in History and Philosophy of
Science, 16 (1985), pp. 171-201 on p. 178.

7. inerton.wikidot.com. Quantum mechanics and de
Broglie's concept. Retrieved September 30, 2011 from
http://inerton.wikidot.com/ quantum-mechanics-and-de-
broglie-s-concept

8. hawking.org.uk. Does God Play Dice? Retrieved
September 30, 2011 from http://www.hawking.org.uk/
index.php/lectures/64

9. Nobelprize.org. Erwin Schrödinger   Biography.
Retrieved September 30, 2011 from
http://www.nobelprize. org/nobel_prizes/physics/
laureates/1933/schrodinger- bio.html

Chapter 10: Energy Bands
No References

Chapter 11: How a Wire Works
No References

Chapter 12: Diodes
No References

Chapter 13: Making a Transistor
No References

Chapter 14: Conclusion
No References

If you enjoyed *Quantum Mechanics A to Z without the BS,* you may enjoy *New Age Quantum Physics.* It goes on to explain quantum mechanics in depth. It answers questions like, "What is the electron cloud?" or "What is energy?" *New Age Quantum Physics* is available from amazon.com and other suppliers of fine books.